THE POETRY OF
COPERNICIUM

The Poetry of Copernicium

Walter the Educator

Silent King Books

SILENT KING BOOKS

SKB

Copyright © 2024 by Walter the Educator

All rights reserved. No part of this book may be reproduced in any manner whatsoever without written permission except in the case of brief quotations embodied in critical articles and reviews.

First Printing, 2024

Disclaimer
This book is a literary work; poems are not about specific persons, locations, situations, and/or circumstances unless mentioned in a historical context. This book is for entertainment and informational purposes only. The author and publisher offer this information without warranties expressed or implied. No matter the grounds, neither the author nor the publisher will be accountable for any losses, injuries, or other damages caused by the reader's use of this book. The use of this book acknowledges an understanding and acceptance of this disclaimer.

"Earning a degree in chemistry changed my life!"
— Walter the Educator

dedicated to all the chemistry lovers, like myself, across the world

COPERNICIUM

Lies Copernicium, in the periodic line.

COPERNICIUM

Its nucleus adorned with protons galore,

COPERNICIUM

Yet fleeting its existence, it yearns for more.

COPERNICIUM

Named after Copernicus, the cosmic sage,

COPERNICIUM

It dances with atoms on history's stage.

COPERNICIUM

In the depths of the cosmos, its tale unfurls,

COPERNICIUM

A fleeting glimpse, in the cosmic swirls.

COPERNICIUM

With electrons arranged in quantum array,

COPERNICIUM

It defies gravity's relentless sway.

COPERNICIUM

In laboratories, it's brought to light,

COPERNICIUM

A fleeting moment, a transient sight.

COPERNICIUM

Its isotopes fleeting, unstable at best,

COPERNICIUM

A fleeting dance, a celestial jest.

COPERNICIUM

With neutrons and protons, it seeks to bind,

COPERNICIUM

In the alchemy of matter, a rare find.

COPERNICIUM

In particle accelerators, it briefly appears,

COPERNICIUM

A fleeting whisper, dispelling fears.

COPERNICIUM

With fleeting half-lives, it fades from view,

COPERNICIUM

In the cosmic dance, forever anew.

COPERNICIUM

Yet in its essence, a mystery lies,

COPERNICIUM

In the depths of the cosmos, where starlight flies.

COPERNICIUM

A trace of Copernicium, in stellar dust,

COPERNICIUM

A fleeting echo, in the cosmic gust.

COPERNICIUM

In the crucible of stars, where elements fuse,

COPERNICIUM

Copernicium emerges, in cosmic hues.

COPERNICIUM

A fleeting moment, in the stellar blaze,

COPERNICIUM

A celestial dance, in the cosmic maze.

COPERNICIUM

So let us ponder, this element rare,

COPERNICIUM

In the tapestry of the cosmos, beyond compare.

COPERNICIUM

A fleeting glimpse, in the grand design,

COPERNICIUM

Copernicium, in the cosmic shrine.

COPERNICIUM

In the laboratory's quest, it's sought with zeal,

COPERNICIUM

A fleeting glimpse, in the cosmic wheel.

COPERNICIUM

With atoms collided, in the particle fray,

COPERNICIUM

Copernicium emerges, if only for a day.

COPERNICIUM

In the annals of science, its story is told,

COPERNICIUM

A fleeting element, elusive and bold.

COPERNICIUM

With protons and neutrons, it seeks its place,

COPERNICIUM

In the cosmic dance, through time and space.

COPERNICIUM

So here's to Copernicium, mysterious and rare,

COPERNICIUM

A fleeting element, beyond compare.

COPERNICIUM

In the symphony of atoms, it plays its part,

COPERNICIUM

A fleeting glimpse, in the cosmic heart.

COPERNICIUM

ABOUT THE CREATOR

Walter the Educator is one of the pseudonyms for Walter Anderson. Formally educated in Chemistry, Business, and Education, he is an educator, an author, a diverse entrepreneur, and he is the son of a disabled war veteran. "Walter the Educator" shares his time between educating and creating. He holds interests and owns several creative projects that entertain, enlighten, enhance, and educate, hoping to inspire and motivate you.

Follow, find new works, and stay up to date with Walter the Educator™ at WaltertheEducator.com

www.ingramcontent.com/pod-product-compliance
Lightning Source LLC
LaVergne TN
LVHW012050070526
838201LV00082B/3882